Light that helps
us to see,

light that grabs attention

and light that warns
of danger.

But humans use clunky equipment and the Earth's resources to do what
some extraordinary animals have evolved to do all by themselves.

In nooks and crannies and under the sea, there are creatures that make their own
light and use it just like us. Curious creatures that really do glow in the dark...

The Sun has set, and two children in Wales
are up late. They run barefoot along the shoreline,
leaving trails of brilliant light in the wet sand.
All around them, waves glow electric blue.

Sea Sparkle

It looks like something from outer space, but this light is
actually made by billions of tiny organisms called *Noctiluca
scintillans* or sea sparkle, floating just below the surface
of the water. These tiny life forms are a bit like animals and
a bit like plants. They drift with the currents of the ocean.

When sea sparkle is disturbed, perhaps by waves, a hungry
predator or the paddle of a kayak, it gives off a vivid light.
It is **bioluminescent**.

Zoë Armstrong

Anja Sušanj

Curious Creatures
GLOWING IN THE DARK

Flying Eye Books

London | Los Angeles

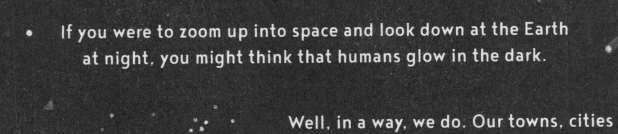

If you were to zoom up into space and look down at the Earth at night, you might think that humans glow in the dark.

Well, in a way, we do. Our towns, cities and motorways glitter with electric light.

Navies are interested in studying sea sparkle. Their boats leave rippling trails of light when they pass through it – not helpful on a stealth mission!

Noctiluca scintillans is the scientific name for sea sparkle. It means 'sparkling night light'.

JUST LIKE YOU...

Sea sparkle might seem very different from humans, but it is communicating just like you do. It uses light as a way of yelling "HELP!" or "GO AWAY!" when it is under threat.

What is Bioluminescence?

The word bioluminescent comes from **bio**, meaning **life**, and **lumen**, meaning **light**. Bioluminescent animals make sparkles, flickers and flashes of light with chemical reactions inside their bodies – a bit like cracking a glowstick. Chemicals called **luciferin** and **luciferase** mix together with oxygen to produce the glow, and it reveals itself in many different ways.

Some creatures have glowing light organs called photophores ...

*The small body of the **firefly squid** is covered in tiny glowing photophores. It uses its bioluminescence to signal to others and to make itself look bigger than it really is.*

... while others ooze bioluminescent slime!

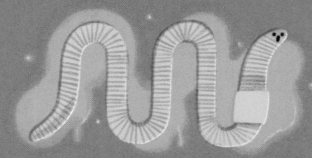

*A New Zealand earthworm called **Octochaetus multiporus** dribbles bioluminescent orange goo when it is disturbed.*

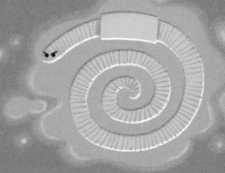

*An earthworm from the American South, **Diplocardia longa**, spews glowing blue gunk to startle its predators.*

Some sea creatures borrow their bioluminescence from glow-in-the-dark bacteria, which they carry around.

Flashlight fish have a glowing torch of bioluminescent bacteria under each eye, which they switch on and off by blinking. They use their torch for finding food and distracting predators with a game of blink-and-run.

Somewhere off the coast of Florida, a marine scientist begins
a deep-sea dive. She wears a special suit that looks more
like a spacesuit than ordinary diving gear.

Down, down, down through salty strands
of seaweed and shoals of colourful fish.
Through pinks and reds, yellows and greens.

The deeper she dives, the darker the ocean
becomes, until no more colours can be seen.
At two hundred and fifty metres below
sea level, she turns out her lamps.

Bioluminescence Under the Sea

Instead of darkness, there is an explosion of
bioluminescent light. Swirls and sparkles and pops
of brilliance everywhere she looks.

Under the sea, more than three-quarters of the
animals glow in the dark. Instead of using colour
to express themselves, they use light.

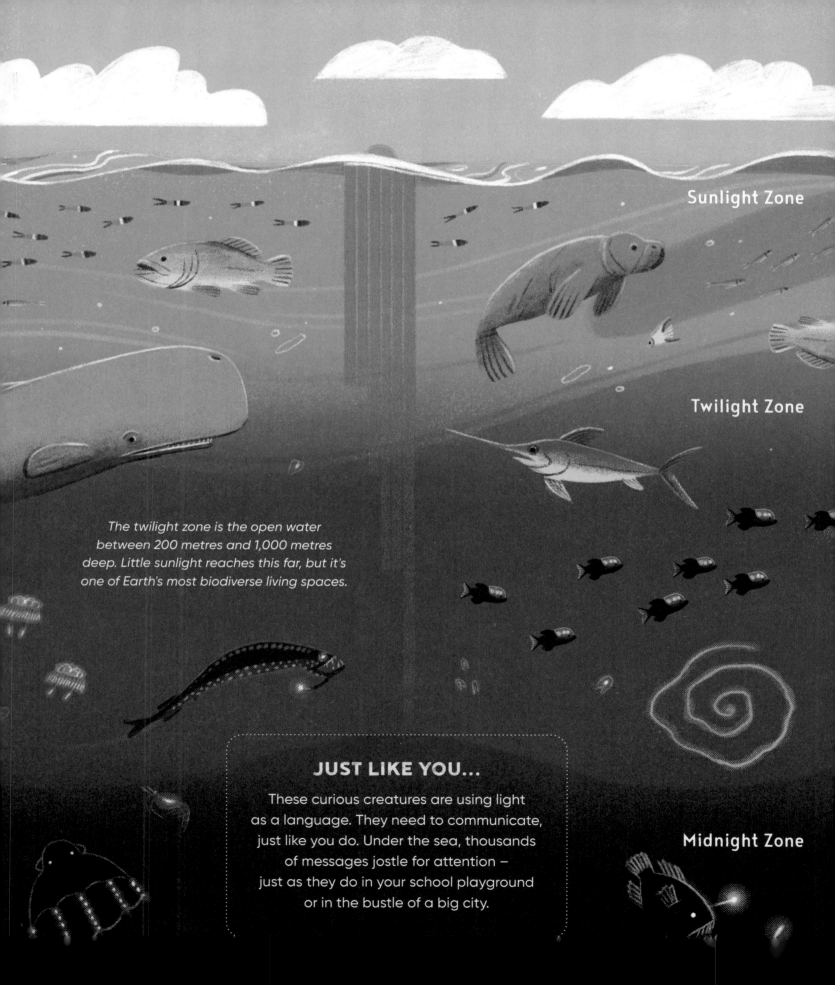

Sunlight Zone

Twilight Zone

The twilight zone is the open water between 200 metres and 1,000 metres deep. Little sunlight reaches this far, but it's one of Earth's most biodiverse living spaces.

JUST LIKE YOU...

These curious creatures are using light as a language. They need to communicate, just like you do. Under the sea, thousands of messages jostle for attention – just as they do in your school playground or in the bustle of a big city.

Midnight Zone

The Meaning of Light

Glowing isn't much use in broad daylight, but in the dark it becomes a superpower.
Bioluminescent animals tend to live in places where the Sun doesn't shine –
deep inside a cave or down in the ocean. Their glow is a way of communicating
through the darkness, helping them to survive and thrive. But what are they saying?

Some creatures want to be noticed ...

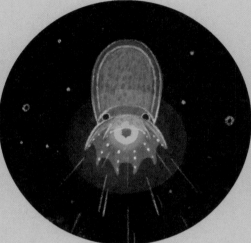

This **bolitaena pygmaea octopus** is searching for a mate.

Her mouth blazes with bioluminescence, as if to say "look at me".

This signal attracts the attention of a male octopus, who shimmies towards her.

... while others want to fit in.

WHOOSH! Hundreds of glittering **headlight lanternfish** dart by, their noses shining like headlights. The light patterns on their bodies may help them swim together with their own species as they rise up to feed. There is safety in numbers and the distinctive patterns say, "Hey, this is our team!"

Sometimes they are spooked ...

There is not much to cling to in the twilight zone, so this **glowing sucker octopus** has evolved useful photophores instead of suckers. A predator approaches and – "BOO!" – the octopus lights up, and startles the creature away.

... and want to stay hidden.

"Shhh!" This **midwater squid** is hiding. Glowing photophores on the underside of its body help it blend in with the brighter water above. To predators lurking below, it is nearly invisible. This is called **counterillumination**.

While some are just hungry.

This **cookiecutter shark** is not much bigger than a cucumber. It uses counterillumination too – and not just for hiding! A dark, non-glowing collar on its 'neck' stands out against the brightness of the water, like the silhouette of a small fish, saying, "Yum! Come and eat me!"

An unsuspecting **spinner dolphin** is lured in and – CRUNCH! – the little shark strikes. It suckers its lips to the side of the bigger creature, taking out a round plug of flesh with its cookie cutter jaws. Dinner!

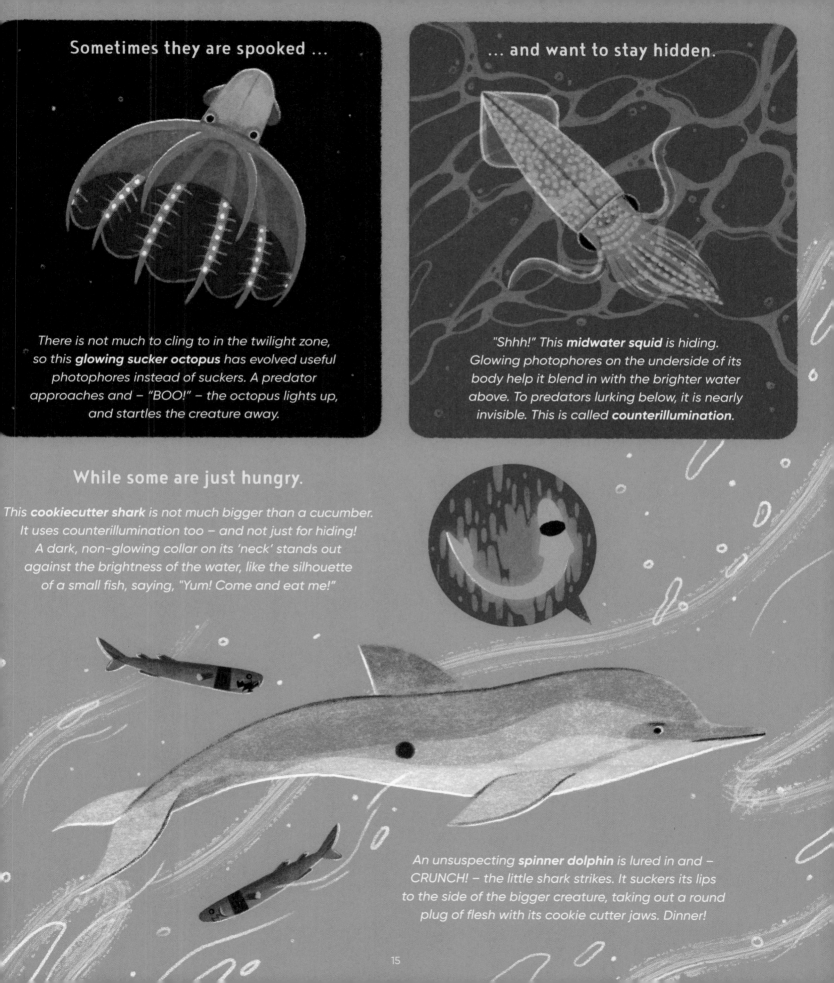

In the murky water of the midnight zone, more
than a thousand metres below the surface of the ocean,
a delicate light drifts slowly through the darkness.

Out of the gloom comes a fearsome
looking fish; she dangles a glowing
lantern over her toothy face.

A passing shrimp is tempted by the light –
it looks like something good to eat. The fearsome
fish opens her huge jaws and CHOMP!
She swallows the shrimp in one gulp.

The Humpback Anglerfish

This curious creature is the humpback anglerfish. Her light comes
from a clump of glow-in-the-dark bacteria, which she carries
on the tip of a long 'fishing rod' protruding from her forehead.

The rod is called the *illicium* and it is actually part of her spine.
The glowing lure on the end is a gland called the *esca*, where the
bioluminescent bacteria is stored.

The crushing weight of the water makes it very difficult for humans to visit the midnight zone. The deep sea is mostly explored using remotely operated vehicles and special underwater cameras.

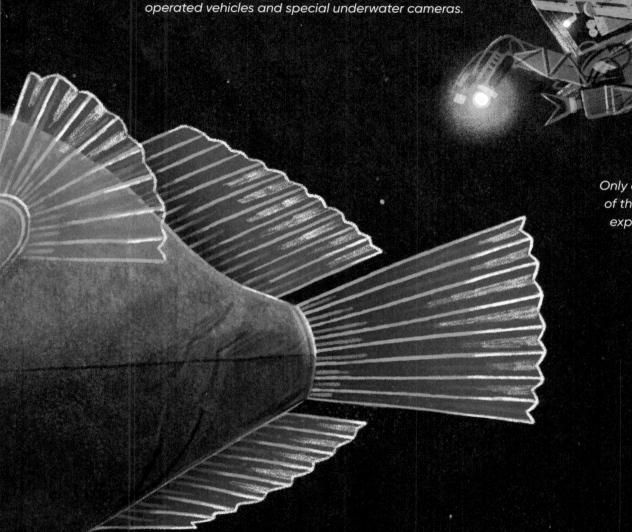

Only about five per cent of the ocean has been explored by humans!

Only the female anglerfish has a light, which is also useful for luring the much tinier males.

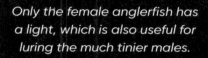

JUST LIKE YOU...

Anglerfish are experts at luring other creatures, just like you might lure a cat with a ball of string, or tempt a puppy with doggie chews. Humans use lures to catch fish too.

Sometimes you might be lured yourself, by the smell of freshly baked cookies or a homemade pie.

Capturing the Light

Around the world and throughout history, humans have tried to harness bioluminescence, and have used it in lots of surprising ways...

Before safety lamps were invented, coal miners in Britain and parts of Europe would light their way through mines with bottles of **fireflies** and jars of glowing **fish skins**. A weak glow was better than no light at all!

During WWII, Japanese soldiers read maps by the light of dried **ostracods** – tiny creatures the size of sesame seeds. The soldiers would add water to make them glow.

In ancient Rome it was said that a walking stick rubbed with the glowing slime of a jellyfish would make a very handy torch.

The Romans were probably talking about the slime of the **mauve stinger**, which leaves a trail of bioluminescent mucus when it is spooked.

A marine scientist called Dr Edie Widder invented an electronic version of a bioluminescent jellyfish, called the 'e-Jelly'. She used it to lure a **giant squid** towards a deep-sea camera. The mysterious squid was longer than a bus, with eyes as big as your head!

Dr Widder based the 'e-Jelly' on the bioluminescent 'scream' of the **Atolla jellyfish**, which flashes like a blue police light when it is attacked. This calls on bigger predators to come over and eat the attacker.

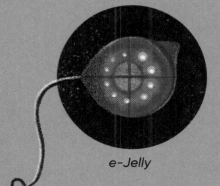

e-Jelly

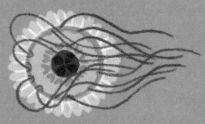

Atolla jellyfish

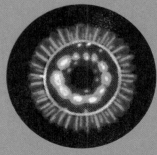

Atolla jellyfish 'scream'

It is a summer's evening in Japan.
A family has gathered at the edge of a forest.
They watch as the Sun sets and sparks
of yellow-green light begin to fill the air.

One by one, the sparks start to flicker as they
float between the trees, until thousands of
tiny lights all flash together in synchrony.

The Japanese Firefly

Look a little closer and you will see that these lights are
made by delicate beetles – a species of Japanese firefly
called *Luciola cruciata*.

Each firefly has a light organ near the tip of its abdomen.
The males synchronise their flashes, while the females blink,
blink, blink to a different beat. This is how the males and
females find each other in the darkness.

Fireflies use their light mainly for courtship, but it has other
functions too. Flashes can warn of danger, and the glow can
say to predators, "Don't eat me. I taste awfully bitter!"

There are more than two thousand species of firefly around the world. The colour of their glow and the sequence of their flashes vary from species to species.

In Japan, firefly watching is a special way to spend a summer's evening. Fireflies shine in poems and songs and films, and it is said that the souls of dead samurai warriors live on in firefly form.

JUST LIKE YOU...

Humans use light to send signals and attract attention too. You might wear light-up running shoes or a glittery jacket to catch someone's eye. Perhaps you have used a torch to flash out coded messages to a friend – this can be fun to try.

The Greatest Light Shows on Earth

Most of Earth's bioluminescent animals live down in the ocean, unseen by humans. But when curious creatures light up on land, they mesmerise us with their dazzling displays. We gather to celebrate bioluminescence – it seems closer to magic than anything we ourselves can make.

*The Māori name for these glowworms is **titiwai**.*

The New Zealand Glowworm

Deep inside a cave in New Zealand, a boat drifts along an underground river. The people on board gaze up at the craggy ceiling, it seems to twinkle with a million stars.

But it's not actually starlight they can see, it's maggots: the glowing larvae of a species of fungus gnat, called the New Zealand glowworm. Visitors come from all over the world to see them.

These glowworms have bioluminescent bottoms,
which they use to catch food in a very peculiar way...

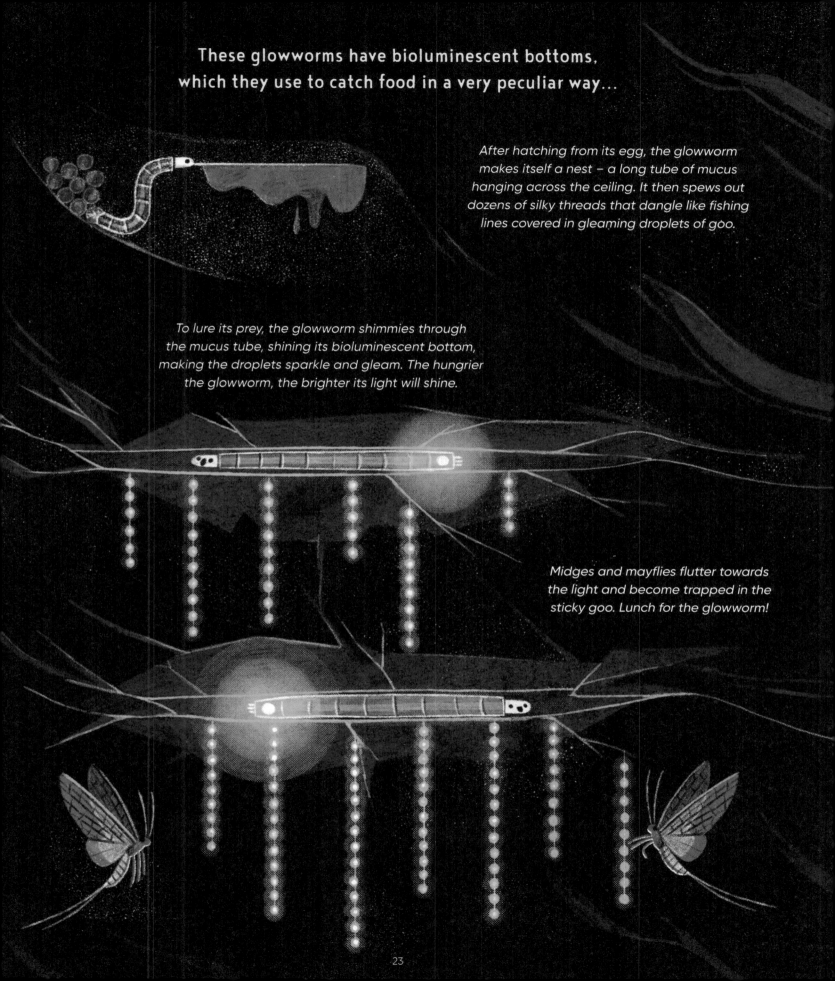

*After hatching from its egg, the glowworm
makes itself a nest – a long tube of mucus
hanging across the ceiling. It then spews out
dozens of silky threads that dangle like fishing
lines covered in gleaming droplets of goo.*

*To lure its prey, the glowworm shimmies through
the mucus tube, shining its bioluminescent bottom,
making the droplets sparkle and gleam. The hungrier
the glowworm, the brighter its light will shine.*

*Midges and mayflies flutter towards
the light and become trapped in the
sticky goo. Lunch for the glowworm!*

The sky over South Africa is growing dark, and these young explorers are searching for a very curious creature. They are using an ultraviolet torch to spot its tell-tale, secret glow.

The ultraviolet light from this torch is invisible to humans. The Sun gives off ultraviolet rays too, as well as the light that we can see.

The Scorpion

Scorpions may look earthy coloured, but that's just what these cunning creatures want you to think. Under the invisible beam of an ultraviolet torch, they glow a brilliant blue-green.

Their glow doesn't come from bioluminescence though. Scorpions have evolved a different way to dazzle – a way that uses energy from the Sun. They are **biofluorescent**.

JUST LIKE YOU...

There are lots of theories about why scorpions glow under ultraviolet light. One idea is that biofluorescence helps to protect them from sunlight, just as you wear sunscreen and seek out shade on a hot day. The glow might warn these night hunters to find cover because it's just too bright out there.

Moonlight is actually sunlight bouncing off the Moon. It contains ultraviolet light too.

Scorpions are nocturnal, meaning they're active at night. They hate light, and even avoid a full moon, probably so predators won't spot them.

What is Biofluorescence?

Unlike bioluminescent creatures, biofluorescent animals only glow when they absorb invisible ultraviolet light from the Sun or the Moon – or a special torch!

Molecules inside the animal transform the ultraviolet light into brilliant shades of blue, green, yellow or red. You might have glow-in-the-dark stars on your bedroom wall that work in a similar way. (Although they glow a little longer after the lights have been turned out.)

A scorpion is a good example of this in action. Its outer shell, called an **exoskeleton**, absorbs the invisible light, then gives off a visible glow. Here's how it works:

Ultraviolet light reaches the surface of the scorpion's exoskeleton.

Molecules in the exoskeleton absorb the ultraviolet light.

The molecules re-emit the light in a different colour, millionths of a second later – this is the glow! When the ultraviolet light is gone, the scorpion stops glowing.

Most land animals have no need to make light – they exist in a sunlit world. But glowing is still a useful trick, especially for nocturnal animals and creatures that live in the dim light of the forest.

These animals don't need a special torch to see the biofluorescent glow – their senses aren't the same as ours. They see colours of light that human eyes can't detect, and the world looks very different...

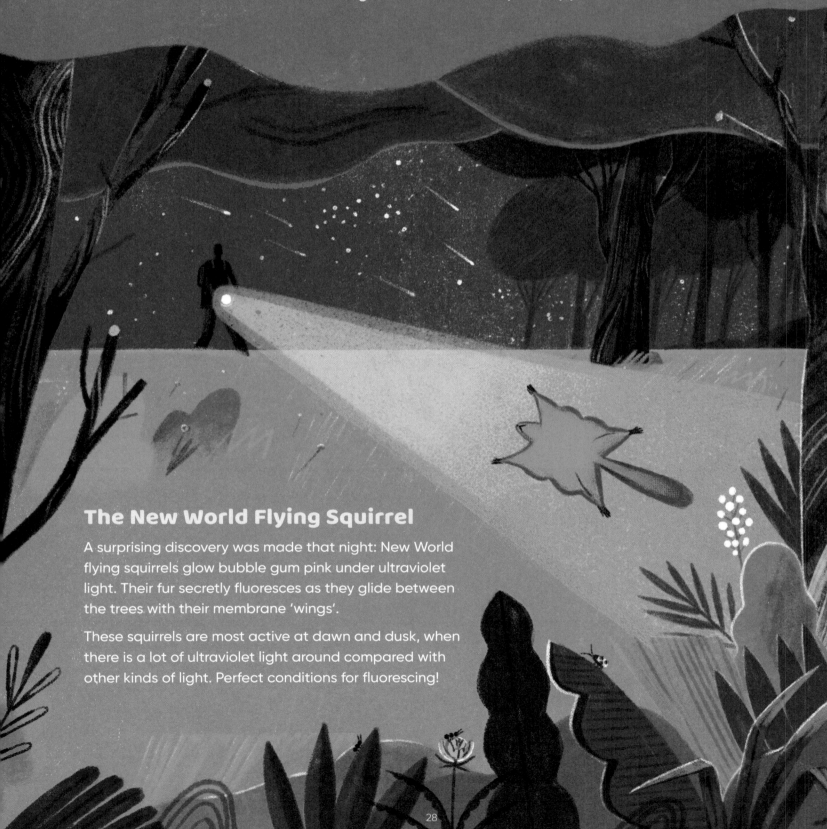

In the American Midwest, a forestry professor is hunting for biofluorescence in the woods. He hears a chirping overhead, and points his ultraviolet torch towards the sound. In amongst the browns and greens, a flash of pink appears.

The New World Flying Squirrel

A surprising discovery was made that night: New World flying squirrels glow bubble gum pink under ultraviolet light. Their fur secretly fluoresces as they glide between the trees with their membrane 'wings'.

These squirrels are most active at dawn and dusk, when there is a lot of ultraviolet light around compared with other kinds of light. Perfect conditions for fluorescing!

Is it a squirrel or an owl? The glow of a flying squirrel might confuse this feathery predator, because owls glow biofluorescent pink too!

Or maybe the glow is camouflage, helping the squirrel blend in with fluorescent lichen or bioluminescent fungus on the trees.

These mushrooms, called Panellus stipticus, are bioluminescent. Several species of fungi glow in the dark!

JUST LIKE YOU...

We don't know for sure why flying squirrels fluoresce. But the pink glow might also help them spot each other in the twilight – just as wearing high-visibility gear makes you easier to spot when you're riding your bike or walking home from school in the dark.

Glowing in Secret

Biofluorescence is something of a mystery – like a language we haven't learned yet. Around the world, scientists are now discovering that all kinds of familiar creatures have been glowing in secret without us even knowing.

The Chameleon

In Madagascar, a chameleon sits in the half-light of the rainforest. Something glows beneath his colourful skin. A pattern of dazzling dots decorates his face and head. The glow is coming from his fluorescent bones.

Ultraviolet light reaches this secret fluorescent layer through 'windows' of thin skin stretched over bumpy bits on the chameleon's skull. Other chameleons can see the dotty, blue glow shining in the forest.

The Polka-Dot Tree Frog

In Argentina, scientists have discovered that this small, green frog is more dazzling than it seems.

The polka-dot tree frog, which lives all over South America, was the first frog in the world found to fluoresce. It gives off a vivid blue-green glow under ultraviolet light. We now know that many other frogs and amphibians are biofluorescent too. Their glow probably helps the creatures to communicate and find a mate.

Scientists have made tiny sunglasses for the puffins, to shield their eyes while their beaks are being studied.

The Platypus

This platypus is a curious creature in every way. It is an Australian egg-laying mammal with webbed feet, a bill like a duck, a tail like a beaver and venomous spurs on its hind feet. And now we learn it is biofluorescent too!

Flabbergasted scientists are trying to figure out why the brown fur of this fantastical river-dwelling animal glows blue-green under ultraviolet light.

The Atlantic Puffin

On a grassy clifftop in Iceland, an Atlantic puffin gulps down some fish. Her patterned beak is colourful and also fluorescent. Under ultraviolet light, the yellow ridges shine brightly as a sign of health, helping the puffin to attract a mate.

Her glow might be useful when feeding puffin chicks too. Pufflings spend weeks in the dim light of an underground burrow. A glowing beak full of fish must be a welcome sight!

A marine scientist is filming biofluorescent coral near the Solomon Islands in the South West Pacific, when a curious creature joins his team. It looks like a flying saucer humming with light as it slips quietly through the water.

The Hawksbill Sea Turtle

This hawksbill sea turtle was the first glowing reptile ever recorded, but the discovery was not as alien as it might seem: many sea creatures that live close to the sunlit surface of the ocean are biofluorescent.

Ultraviolet and blue light pass through the shallow waters, allowing animals such as this turtle to dazzle.

Perhaps the green and red of its biofluorescent shell help to camouflage the turtle as it forages for food along the glowing coral reef. Blending in makes it harder for predators to spot.

Scientists have discovered biofluorescence in hundreds of species of fish, and they are just getting started!

There are very few hawksbill sea turtles left in the oceans – the species is critically endangered – so the reasons for their glow might always be a mystery.

JUST LIKE YOU...

The shy hawksbill sea turtle blends in against the rocks and glowing coral, just as you may like to blend in or hide sometimes too. You might play hide-and-seek, or dress to match your friends. It's nice to be different but sometimes blending in makes us feel safe – just like the turtle.

A Glowing Future

Creatures that glow in the dark are inspiring humans to find extraordinary new ways to save energy, protect the planet and track diseases.

Bioluminescence can help us to look after our oceans. In Florida, glow-in-the-dark bacteria are being used to test for pollution. A bright glow means healthy water, but dim light means pollution by pesticides, heavy metals and plastics.

Medicine borrows from nature too. Bioluminescence and biofluorescence can help scientists track how diseases move around the body. This helps us understand how to make people better.

A team in the United States created glowing watercress, using the same chemicals that give bioluminescent animals their light. Now scientists in Russia, the UK and Austria have made glowing plants using genes from bioluminescent mushrooms. The idea is that one day we might save energy with living desk lamps that don't need plugging in!

Scientists in Denmark are trying to invent glow-in-the-dark trees, which they dream might one day help to light our cities!

We share our planet with these extraordinary animals that glow in the dark. It's easy to see our differences, but we also have lots in common. They are driven to communicate...

to survive ...

and to thrive ...

just like you.

These creatures have been making light
on Earth for millions of years. Now it is up
to us to keep their light glowing.